Edison Twp. Free Public Library
340 Plainfield Ave.
Edison, New Jersey 08817

My First Science Words

WEATHER WORDS

A CRABTREE SEEDLINGS BOOK

Taylor Farley

CRABTREE
PUBLISHING COMPANY
WWW.CRABTREEBOOKS.COM

weather
(WETH-ur)

sunny
(SUHN-ee)

cloudy
(CLOU-dee)

rainy
(RAYN-ee)

stormy
(STOR-mee)

windy

(WIN-dee)

tornado
(tor-NAY-doh)

snowy
(SNOH-ee)

foggy
(FOG-ee)

rainbow
(RAYN-boh)

Glossary

cloudy (**CLOU-dee**): Cloudy weather has so many clouds in the sky, the Sun cannot shine through.

foggy (**FOG-ee**): Foggy weather makes it hard for us to see outside. Fog is made of water in the form of a gas which you cannot see.

rainbow (**RAYN-boh**): A rainbow is an arc of different colors. Rainbows appear when the Sun shines a certain way through water droplets in the air.

rainy (**RAYN-ee**): Rainy weather happens when rain falls from clouds.

snowy (**SNOH-ee**): Snowy weather happens when snow falls from clouds.

stormy (STOR-mee): Stormy weather has lots and lots of rain or snow. It may also be very windy and have thunder and lightning.

sunny (SUHN-ee): Sunny skies have few or no clouds to block the light from the Sun.

tornado (tor-NAY-doh): A tornado is a powerful, spinning column of air shaped like a funnel. Tornados are dangerous.

weather (WETH-ur): Weather describes what the air is like outside (hot, cool...). It also describes how the air is moving (wind) and what it is carrying (rain, snow...).

windy (WIN-dee): Windy weather happens when the air outside is moving fast.

School-to-Home Support for Caregivers and Teachers

Crabtree Seedlings books help children grow by letting them practice reading. Here are a few guiding questions to help the reader build his or her comprehension skills. Possible answers are included.

Before Reading
- **What do I think this book is about?** I think this book is about weather. It might tell us about different kinds of weather.
- **What do I want to learn about this topic?** I want to learn about the clothing people wear in different weather.

During Reading
- **I wonder why...** I wonder why it looks like there are two rainbows in the picture on page 21.

- **What have I learned so far?** I have learned that people wear bathing suits in sunny weather and jackets in cloudy weather. They wear hats, coats, and mittens in snowy weather.

After Reading
- **What details did I learn about this topic?** I learned that it can be difficult to drive in foggy weather. Cars have lights to help people see.
- **Write down unfamiliar words and ask questions to help understand their meaning.** I see the word *rainy* on page 8 and the word *tornado* on page 14. The other vocabulary words are found on pages 22 and 23.

Library and Archives Canada Cataloguing in Publication

Title: Weather words / Taylor Farley.
Names: Farley, Taylor, author.
Description: Series statement: My first science words | "A Crabtree seedlings book". | Previously published in electronic format by Blue Door Education in 2020.
Identifiers: Canadiana 20200384910 | ISBN 9781427130471 (hardcover) | ISBN 9781427130525 (softcover)
Subjects: LCSH: Weather—Terminology—Juvenile literature. | LCSH: Climatology—Terminology—Juvenile literature.
Classification: LCC QC981.3 .F37 2021 | DDC j551.6—dc23

Library of Congress Cataloging-in-Publication Data

Names: Farley, Taylor, author.
Title: Weather words / Taylor Farley.
Description: New York, NY : Crabtree Publishing, 2021. | Series: My first science words ; a Crabtree seedlings book | Audience: Ages 4-6 | Audience: Grades K-1 | Summary: "This book builds beginning vocabulary about the science of weather. Extremely helpful for elementary science preparation, eight words combine with a visual depiction so readers can see what the word means"-- Provided by publisher.
Identifiers: LCCN 2020049632 | ISBN 9781427130471 (hardcover) | ISBN 9781427130525 (paperback)
Subjects: LCSH: Weather--Juvenile literature.
Classification: LCC QC981.3 .F33 2021 | DDC 551.5--dc23
LC record available at https://lccn.loc.gov/2020049632

Crabtree Publishing Company
www.crabtreebooks.com   1–800–387–7650

Written by Taylor Farley
Production coordinator and Prepress technician: Samara Parent
Print coordinator: Katherine Berti

e-book ISBN 978-1-947632-70-7

Print book version produced jointly with Blue Door Education in 2021

Printed in the U.S.A./012021/CG20201102

Content produced and published by Blue Door Publishing LLC dba Blue Door Education, Melbourne Beach FL USA. Copyright Blue Door Publishing LLC. All rights reserved. No part of this book may be reproduced or utilized in any form or by any means, electronic or mechanical including photocopying, recording, or by any information storage and retrieval system without permission in writing from the publisher.

Photo credits: page 2 © Shutterstock.com/Thomas Amby; page 3 © Shutterstock.com/solarseven, weather symbols © Shutterstock.com/ En min Shen; cover and page 5 © Shutterstock.com/ Anton Sterkhov; page 7 © Shutterstock.com/ Lilly Trott; cover and page 9 © Shutterstock.com/ A3pfamily; page 11 © Shutterstock.com/ page 11 © Shutterstock.com/ Studio 1One; page 13 © Shutterstock.com/ Neil Lockhart; cover and page 21 © Shutterstock.com/ Tomsickova Tatyana; page 23 © Shutterstock.com/ Kichigin; rainbow icon © Shutterstock.com/Solaie; page 19 © Shutterstock.com/bogdan ionescu ©shutterstock.com/Brian A Jackson www.Shutterstock.com

Published in Canada
Crabtree Publishing
616 Welland Ave.
St. Catharines, Ontario
L2M 5V6

Published in the United States
Crabtree Publishing
347 Fifth Ave.
Suite 1402-145
New York, NY 10016

Published in the United Kingdom
Crabtree Publishing
Maritime House
Basin Road North, Hove
BN41 1WR

Published in Australia
Crabtree Publishing
Unit 3 – 5 Currumbin Court
Capalaba
QLD 4157

Edison Twp. Free Public Library
340 Plainfield Ave.
Edison, New Jersey 08817